YOUR KNOWLEDGE HAS VALUE

- We will publish your bachelor's and master's thesis, essays and papers

- Your own eBook and book - sold worldwide in all relevant shops

- Earn money with each sale

Upload your text at www.GRIN.com and publish for free

Mohmed Sarhan

Use of legume grains in organic animal feeding. Effect on sheep milk yield and quality

GRIN Publishing

Bibliographic information published by the German National Library:

The German National Library lists this publication in the National Bibliography;
detailed bibliographic data are available on the Internet at http://dnb.dnb.de .

Imprint:

Copyright © 2011 GRIN Verlag GmbH
Print and binding: Books on Demand GmbH, Norderstedt Germany
ISBN: 978-3-656-88961-8

Università degli Studi di Palermo

Facoltà di Agraria di Palermo

Dipartimento DEMETRA

Course of Laurea in Organic farming and quality management

Use of legume grains in organic animal feeding: effect on sheep milk yield and quality

Tesi di Laurea di:
Mohamed Sarhan

Relatore:
Dott. Massimo Todaro

Correlatore:
Ch.mo Prof. Paolo Inglese

Anno Accademico 2010-2011

*Abstract**

In this context fits the current need to substitute in animal feed, soybeans and corn for food alternative, depending not only of their high market price, but also the risks associated with GMOs and the presence of possible contamination mycotoxin.

As for corn, an energy source that most commonly found in commercial foods, in Mediterranean environments and animals with high nutritional requirements does not, it is replaced with barley, corn can find it easily in Sicily. As for the replacement of soybean, grain legumes, traditionally cultivated in Sicily, respond well to this need, both for their willingness at the local level, both being GM - free and therefore more nutritionally safe.

The objective of this thesis is therefore to provide a contribution to the knowledge of the chemical-nutritional few grains of legumes grown locally to be used as protein sources for feeding of sheep as an alternative to soy milk, and get directions on the legume species in order to pursue, along with good production levels, better diet and organoleptic characteristics, as well as a high security level nutrition of dairy sheep. To this end, we evaluated the effects of the food administration of chickpea, Faba bean or pea seeds, mixed with barley, production and quality of milk obtained from lactating ewes.

*Corresponding author: mohmedsarhan@gmail.com

Introduction

1. Sustainability of Ruminant Production system

Sustainability of ruminant agriculture is the overall result of the interaction among multi-dynamic functions (fig1). Ruminant production sustainability constitutes a hot topic that has been the subject of numerous publications and conflicting reports in recent decades (Delgado et al., 1999; Steinfeld et al., 2006; The World Bank, 2009).

That is warranting a rational progress at all levels (farm, local, regional, national and international) without compromising environmental stability (i.e. avoiding soil erosion, desertification or greenhouse gases emissions (*GHG*)), whereas contributing to food security and poverty alleviation programmes.

Rational feeding and nutrition systems are essential in this goal. A more holistic approach of research targets is required in which physiological functions and farmers' practices must converge and respond to each particular situation in an integral, dynamic and flexible conceptual perspective.

Competition for feed/food use is still a crucial criterion, future ruminant feeding systems (*FeSyst*) should preferably focus on lignocellulosic sources according to biome distributions, and the recent increases in volumes of crop residues and their by-products, the annually renewed volumes of these biomasses are considerable.

There are numerous recent studies highlighting sustainability problems for the development of ruminant production systems (*RPS*) while facing increasing human food necessities and global climate change.

The main objectives of the ruminants' physiologist should be convergent for both industrialized (*IC*) and developing countries (*DC*) in a common and global strategy of advancing knowledge. In DC, this means improving the efficiency of RPS, taking into account the unique possibility of using rangelands. For IC settings, RPS should be revisited in terms of autonomy and environment-friendly feeding and managing practices. (Delgado 2005).

The renewed vision of ruminal digestion through the reduction of greenhouse gas emissions is also a key aspect as it is an environmental demand that cannot be ignored. With regard to other ruminants' physiological functions, accumulated knowledge could be mobilized into an integrative approach that puts forward the adaptive capacities of animals to face variability in quantity and quality of supplied feeds.

Exploiting traditional techniques of forage conservation (hays, silages) will continue to be strategic to alleviating seasonal effects (e.g. drought and winters) and optimizing forage use at the farm level (Nussio and Ribeiro, 2009; Orosz et al., 2009).

Grains and high-energy feeds. Classically, intensive livestock production 'laws' dictate that, 'to maximize production', the livestock producer must use a high-energy ration that is, a high caloric density and digestibility (e.g. low fibre content). In grain-producing countries (e.g. ICs), a great part of the livestock energy requirement has been traditionally warranted by local cereal production (e.g. maize, wheat, barley, rice, oat and sorghum) (FAO, 2009b).

Rising maize and soya bean prices due to agro fuel production makes it more difficult to use them in animal production from a stable and sustainable perspective. The immediate solution for the livestock sector is to Centre attention on the possibilities of becoming less dependent on cereals and oil-seeds in their FeSyst strategies.

Livestock keepers are aware, and therefore, strategies such as using energetic by-products, are becoming more important (FAO, 2008b).

By-products generate credits to the biofuel production chain and thereby greatly improve the energy and environmental performance. When by-products are used for heat and process energy the energy balance improves. Credits for by-products are an important element in the calculation of greenhouse gas (GHG) reductions of the different biofuel production chains compared to fossil fuel use (Edwards's et.al 2006).

Recently several studies have attempted to include attribution of those by-products, which can be used as animal feed in an analysis of land use requirements of biofuels (Özdemir et.al, 2009; Gallagher Review, 2008).

2. Legumes

Mediterranean legume grains have a considerable role as protein suppliers in animal diets. However, since the late 1960s their importance has been reduced due mainly to the use of oleaginous meals, mostly soybean which have been available at competitive prices. As a consequence, an important reduction in legume grain cultivation was noted (Cubero, 1983). This was due to a lack of genetic and technological improvement. This seems to have happened also in other countries in South Europe (Duffus and Duffus, 1991).

In order to reduce the dependence of Europe in imported legume grains as protein sources for animals, the EU has initiated some production incentive policies (Todorov, 1988; Chominot, 1992).Within the few researchers who have studied the nutritive value of mediterranean legume grains, special reference must be made to the work of Wiseman and Cole (1988) with pigs and Castanon and Perez-Lanzac (1990) with layer hens, and of Hadji- panayiotou et al. (1985) with sheep, pointed to the possibility of using these grains in animal diets, although in limited amounts. Legumes are more than just peas and beans because they comprise the third largest group of flowering plants in the world with more than 18,000 species. Found in environments as diverse as the arctic tundra, dry scrublands and tropical rainforest. They may be trees, shrubs, vines or herbs.

Legumes Defined by the structure of their fruits as all legumes have seed pods that are seamed on two sides. The seeds are arranged in a single row and attached to the pod along one seam. The pods are extremely variable, ranging from small single-seeded types to woody pods a meter long. The pods of most legumes are dehiscent they split along both seams when ripe. Others have indehiscent pods called loments that are constricted between the seeds.

These split into single seed segments when ripe. Still others have a single seed in a winged pod and are dispersed by wind. Many legumes are used extensively as shade trees or for ornamental plants because of their attractive foliage or flowers.

Grain legumes crops represent a great resource in organic agriculture both to satisfy the nutritional content of organic livestock feeding. The taxonomic classification of legumes is the subject of ongoing heated debate. Formerly known as   **Leguminosae**, most taxonomists currently classify them as the family **Fabaceae**, which can be further divided into three sub-families – Faboideae, Mimosoideae and Caesalpinioideae. Others consider these as three separate but closely related families, calling them Fabaceae, Mimosaceae and Caesalpiniaceous. The sub-families (or families) are based on the form of the flowers, specifically the petal shape:

Faboideae

Showy flower has one large creased petal (called the banner), (The wings), and two bottom petals joined forming a boat-shaped structure (The keel).Pollinated by bees and the flower contains 10 fused stamens. The faboideae is the largest sub-family, most of which are herbs or small Shrubs. Includes many of the legumes such as beans, alfalfa and Peanuts.

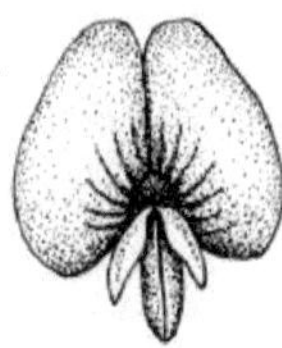

Faboideae flower

Mimosoideae

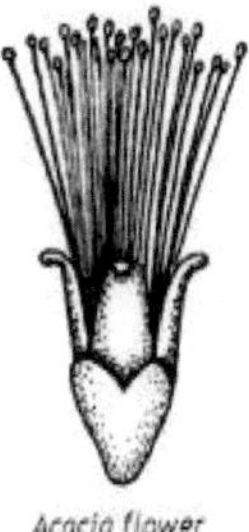

Acacia flower

The flowers of this sub-family are actinomorphic (they can be divided into two symmetrical halves by a line drawn anywhere straight through the centre). They have small petals and the stamens are usually the showiest part of flower. Most of the 2,800 species worldwide are small trees and shrubs found in semi-arid tropical regions of Africa, North and South America and Australia. The Acacia species, which include the Australian wattles, are the most well-known members of the group.

Caesalpinioideae

This sub-family is distinguished by flowers having five large, equal-size petals.It is a diverse group that probably diverged from the other sub-families early in their evolutionary history and lacks many of the classic floral features of other legumes. Of the approximately 2,500 species worldwide, most are trees that are native to tropical savannahs in Africa, South America and Asia. Many of these have become important ornamental plants.

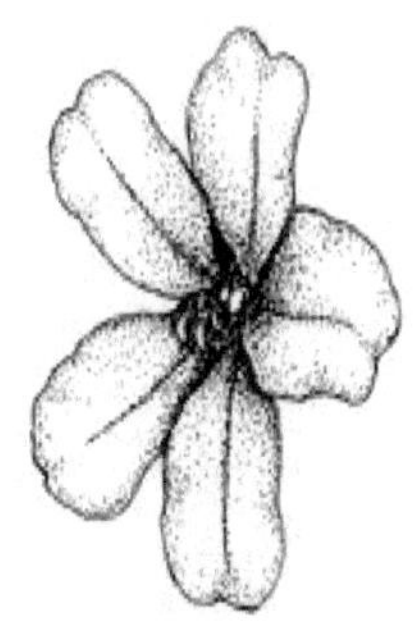

Senna roemeriana flower

3. Nitrogen Fixation by Legumes

Working with a group of bacteria called rhizobia, legumes are able to pull nitrogen out of the air and accumulate it biologically.

The bacteria, which are normally free-living in the soil in native range of aparticular legume, infect (inoculate) the root hairs of the plant and are housed in small root structures called nodules. Energy is provided by the plant to feed the bacteria and fuel the nitrogen fixation process.

In return, the plant receives nitrogen for growth. There are thousands of strains of rhizobia. Certain of these will infect many hosts, certain hosts will accept many different strains of rhizobia. Certain hosts may be nodulated by several strains of rhizobia, but growth may be enhanced only by particular strains. Therefore, when introducing hosts to a new area it is extremely important to also introduce a known effective symbiotic rhizobia strain.

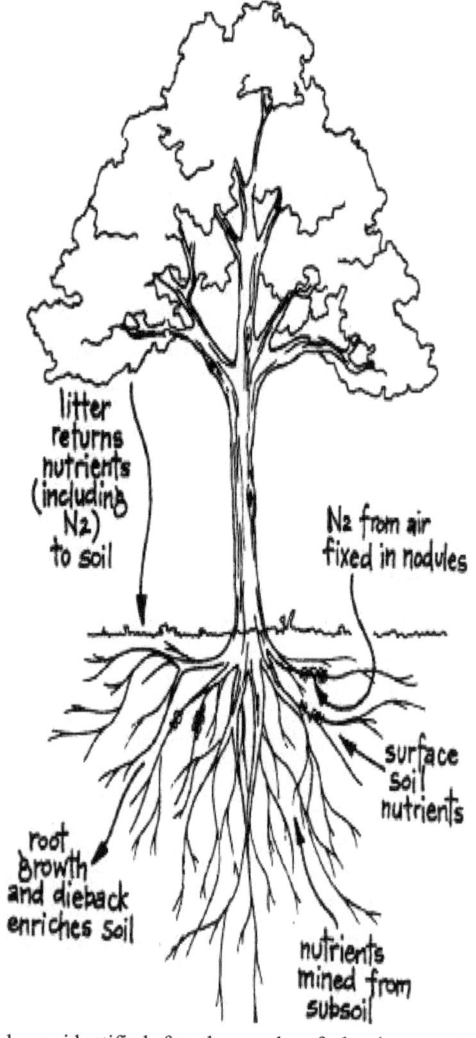

Such effective strains have been identified for thousands of the important nitrogen fixing legumes.It is estimated that 50-800 kg of nitrogen per hectare per year are accumulated by nitrogen fixing plants, depending on species, soil and climate, rhizobium effeciency, and methodology. This biologically fixed nitrogen is an alternative source to inorganic nitrogen fertilizer for crop growth.

Major grain legumes are estimated to fix approximately 11,1 million metric tons of nitrogen per annum from the atmosphere in developing countries as measured by stable isotop methodologies.

Legume nitrogen fixation starts with the formation of a nodule. A common soil bacterium, Rhizobium, invades the root and multiplies within the cortex cells.

The plant supplies all the necessary nutrients and energy for the bacteria. Within a week after infection, small nodules are visible with the naked eye. In the field, small nodules can be seen 2-3 weeks after planting, depending on legume species and germination conditions. When nodule are very young and not yet fixing nitrogen, they are usually white or gray inside.

As nodules grow in size, they gradually turn pink or reddish in color, indicating nitrogen fixation has started. The pink or red color is caused by leghemoglobin (similar to hemoglobin in blood) that controls oxygen flow to the bacteria.The nitrogen fixed is not free. The plant must contribute a significant amount of energy in the form of photosynthate (photosynthesis derived sugars) and other nutritional factors for the bacteria.

Nitrogen fixation occurs in nodules on legume roots (Source: FAO, Rome)

A soybean plant may divert 20-30 percent of its photosynthate to the nodule instead of to other plant functions when the nodule is actively fixing nitrogen.

The stalks, leaves and roots of grain legumes, such as soybeans and beans contain about the same concentration of nitrogen as found in non-legume crop residue.

Infact, the residue from a corn crop contains more nitrogen than the residue from a bean crop, simply because the corn crop has more residue.

A perennial or forage legume crop only adds significant nitrogen for the following crop if the entire biomass (stems, leaves, roots) is incorporated into the soil.

If a forage is cut and removed from the field, most of the nitrogen fixed by the forage is removed. Roots and crowns add little soil nitrogen compared with the aboveground biomass. When the grain from a grain legume crop is harvested, little nitrogen is returned for the following crop. Most of the nitrogen fixed during the season is removed from the field.

4. Grain Legume Crops Use in Animal Feeding For Organic Production

Grain legumes crops represent a great resource in organic agriculture both to satisfy the nutritional content of organic livestock feeding and to maintain soil fertility.

The commercial availability of organic grain legume is decreasing, the costs are high and the GMO contamination risk is particularly high for soya bean, used to achieve the high protein values required by the animals.

So, the cultivation of grain legumes such as sweet lupin (Lupinus albus), field bean (Vicia faba var. minor), and high protein pea (Pisum sativum) on farm could solve the problem and improve the sustainability of the farm.

The use of legumes as source of protein for the animal feed industry is expected to increase further in the near future. Rising incomes in the Asian region are increasing the demand for meat products, and hence the requirement for animal feeds.

There have been changes in public perception and some unfortunate developments, such as the consequences of 'mad cow' disease (i.e. bovine spongiform encephalopathy or BSE) in UK. This has resulted in many feed compounders either choosing to, or being banned from, using animal by-products as a source of protein.

The amino acid requirement of animals often differ with species and bodyweight, hence no single source of plant protein will provide the exact amino acids required for all animals. It is, therefore, preferable to include a range of protein sources in diet formulations, each complementing the other, *(Franco Jubete, 1991).*

For these reasons, the demand for grain legumes, such as field peas (*Pisum sativum*) and faba beans (*Vicia Faba),* by the feed industry is expected to increase. Any feedstuff is likely to be used in diets for animals if it supplies the required nutrients, if it is cost competitive with other available ingredients, and if the user is confident it will produce the desired result. The possibility to substitute external soya bean, a high risk GMO (genetically modified crops) alimentary source, with other legumes produced on farm, such as sweet lupin, protein pea and field bean, as alternative protein source in the formulation of diet in organic dairy cattle nutrition because Grain legumes crops represent a great resource in organic agriculture both to satisfy the nutritional content of organic livestock feeding and to maintain soil fertility.

The commercial availability of organic grain legume is decreasing, the costs are high and the GMO contamination risk is particularly high for soya bean, used to achieve the high protein values required by the animals. So, the cultivation of grain legumes such as sweet lupin (*Lupinus albus*), field bean (*Vicia faba),* high protein pea (*Pisum sativum*) on farm could solve the problem and improve the sustainability of the organic farm. Because the climate (often too dry, too hot and too cold) and soil characteristics (rocky soils).

The Mediterranean area normally offers poor pastures and scarce possibilities to produce a sufficient amount of vegetal protein sources to feed dairy cattle (Boyazoglu and Morand, 2000). But, alternative of buying soya bean is feasible in many countries, like in the Central Italy. In fact the production on farm level of field bean, protein pea and sweet lupine provide considerable contributions of nutrient for cattle. Protein pea is very interesting for its higher grain yield while field beans may suffer under climatic condition resulting in lower yields.

Anyway more research work is needed on the use of local varieties. The introduction of lupine as alternative to soya bean in the formulation of diet in organic dairy cattle nutrition is interesting also in case of bitter lupine. In fact, when mixed with field bean and protein pea, to make it more palatable, the quantity and quality productive levels are slightly lower in comparison with soya bean diet.

Production of food from animal origin in Mediterranean countries was always constrained by the insufficient and fluctuating natural supply of feeds (Molina-Alcaide et al., 2003).

The quality of legumes can be assessed by the type and quantity of fibrous material in the plant. Neutral detergent fiber (NDF) is the cell wall material of the plant and is comprised of the hemicellulose, cellulose, and lignin. Increasing levels of NDF in plants and/or diets have been found to limit dry matter intake (DMI). Acid detergent fiber (ADF) is comprised of cellulose and lignin and is closely associated with digestibility.

An increase in the indigestible lignin complex in the ADF fraction reduces digestibility of the plant. Therefore, ADF and digestibility are negatively correlated. Protein content of a legume is important, but it does not affect the two most important quality characteristics intake and digestibility.

Quality of legumes is best estimated by their potential DMI and dry matter digestibility (DDM) which is determined by the NDF and ADF fractions, respectively. Both NDF and ADF increase as the plant matures causing a decline in the quality of the forage (Jim Linn and Carla Kuehn 1987).

The European Mediterranean areas are suitable for growing grain legumes, which can potentially contribute to increased production of traditional Mediterranean animal products and have positive effects on food safety. Also, legumes could improve soil quality in non-irrigated lands by increasing nitrogen-fixing bacteria (Graham and Vance, 2003).

The benefit of any feed protein in ruminants depends on the protein digestion in the small intestine (Calsamiglia and Stern, 1995; McNiven et al., 2002).

4.1. *Non ruminant animal*

Both the leaf and seed of legumes can be used in animal feed, but in the monogastric animal the seed is more important than the leaf. Legume seeds are an important source of a plant protein in human and animal diets. Many countries are increasing their production of seed legumes (Lu and Gangyi, 2008).

In Asia, about 20 legume species, most of them tropical, are cultivated, mainly for human consumption (Ravindran & Blair 1992).

Raw legume seeds are important source of protein and other nutrients for monogastric animals. However, these legume seeds include many kinds of anti-nutritive factors (ANF) such as trypsin and chymotrypsin inhibitors, lectins and tannins. Although these ANF vary by species, cultivation areas and harvesting methods, they play a role in decreasing the utilization of nutrients (Gatel 1994).

Supplementing antibiotics in swine and poultry diets at subtherapeutic levels has long been shown to increase the growth rate (Becker et al., 1955; Stahly et al., 1980; Cromwell, 2002). However, due to growing concerns over antibiotic resistance, the use of antibiotics as growth promoters in animal diets could be limited due to public or regulatory pressures. Thus, current research involving feed additives for diets of young animals is focused on searching for alternatives to antibiotics that would have at least similar growth-promoting effects of antibiotics without causing bacterial resistance.

Phytobiotics is a term used to describe plant-derived natural bioactive compounds, which affect animal growth and health, and is often applied to essential oils, botanicals, and extracts derived from herbal plants. Some phytobiotics are known to have antimicrobial or antiviral activities (Sökmen et al., 2004; Piao et al., 2006; Friedman, 2007).

Those selected herbs or phytobiotics have long been used as complementary or alternative medicine to improve human health or to cure human disease. Recent advances in science have allowed for the identification of active components from selected phytobiotics and investigation into the mechanisms of those components in the animal's body.

More recently, animal nutritionists have attempted to use some phytobiotics as alternatives to the use of antibiotics for young animals and birds (Mao et al., 2005; Kommera et al., 2006; Peeters et al., 2006; Yuan et al., 2006). Seed rejected for consumption by humans is available for livestock feed.

4.1.1. pig

The pig is a monogastric animal which is slightly tolerant to the ANF in raw legume seed. Nevertheless, newly weaned piglets and young poultry are highly susceptible to various stressors, such as bacterial diseases resulting in decreased growth and even death (Lecce, 1986; Hoerr, 1998).

The optimal use of legume seed as a pig feed necessitates a lowering of the toxicity in the raw seed until it is appropriate for incorporation in the feed. There are many methods to improve the utilization of nutrients in legume seed, such as breeding improvement, physical treatments (decortications, dehulling, milling and others), heat treatments (toasting, boiling, extrusion, streaming or autoclave), chelating substances for binding toxics, radiation and soaking. The choice of the treatment depends on the availability of facilities and economic considerations.

The potential utilization of legume seeds as a source of protein and energy for pigs is, however, governed not only by their essential amino acid and digestible energy (DE) content, but also by the possible presence of (ANF). Therefore, the extent to which legumes are used in pig production is still limited because of problems with their effective nutritional quality.

4.2. Ruminant animal

4.2.1. Cow

Milk production may decrease in early lactation when the demand for rumen undegradable protein is high. The reduction in milk production is attributed to the greater degradation of pea protein in the rumen (Khasan et al., 1989).

Field peas contain high levels of protein (approximately 24.5 % CP) and energy (approximately 88% TDN), and are an attractive, nutrient dense livestock feed for ruminants and non-ruminants.

Since field peas are handled as a dry grain legume for livestock feed, they can be stored and processed in a manner similar to other grains.

Field peas have been proven to be an attractive ingredient in creep feeds with increased feed intake and gain with increasing inclusion of peas up to 67% of the diet (Anderson, 1999; Landblom and Poland, 2000).

Whole grains are usually less digestible than processed grain. Dry rolling has a tendency to split the pea into the hull fraction and the endosperm fraction. The lighter hull fraction may not mix well and possibly adds variation to the diet. The particle size of ground peas is smaller resulting in increased surface area for enzyme and microbial activity.

The milk and blood urea increase in dairy cow when feeding flaked because of an increase of ammonia in the rumen due to a lack of effect of steam-flacking on pea protein degradability.

4.2.2 Buffalo

In the organic buffalo (Bubalus bubalis) dairy farms the protein needs are mainly supplied by soybean (heat-treated seed and cake). Nevertheless, in the Mediterranean regions this oilseed is often difficult to cultivate in organic systems and this contrasts with the one of the principles of organic farming that is to produce all feeds in site. Grain legumes, e.g. pea (Pisum sativum), faba beans (Vicia faba) and lupines (Lupinus spp.) can complete cereals in animal feed and are well adapted to Mediterranean growth conditions. Besides being a valuable protein source, these grain legumes benefit the farming system via biological nitrogen fixation. Peas, in particular, are a high quality and relatively cheap source of protein, used in monogastrics and ruminant feed (Moschini et al., 2008; Masoero et al., 2008).

4.2.3 Small ruminant

Sheep and goats are kept for both meat and milk; small ruminants require energy, protein, vitamins, minerals, fiber, and water. Energy (calories) is usually the most limiting nutrient, whereas protein is the most expensive.

Deficiencies, excesses, and imbalances of vitamins and minerals can limit animal performance and lead to various health problems. Fiber (bulk) is necessary to maintain a healthy rumen environment and prevent digestive upsets. Water is the cheapest feed ingredient, yet often the most neglected. (Gatel 1994).

Secondary metabolites such as tannins and tanniferous plants have implications in organic goat production (Puchala et al., 2008).

Major considerations of organic goat production include personnel, livestock (animal welfare and health), facility, management, environmental conservation, product quality and safety, and profitability. Essential elements for a successful organic goat production may include nutrition and feeding, reproduction, breeding, animal health, animal welfare, environment, plants and vegetation and workers, therefore understanding of nutrition, eating behavior and seasonal interaction can also be beneficial to improve production efficiency. (Lu and Gangyi, 2008).

It has been demonstrated that protein supplementation can improve resistance and/or resilience of grazing goats (Chartier etal., 2000; Eitter et al., 2000; Wang et al., 2008).

In a survey (40 variables in 10 indicators) of 23 farms (organic and conventional) in Andalusia marginal land areas, Mena et al. (2008) reported that conventional goat production systems lag behind organic in nutritional management, sustainable pasture management, soil fertility and contamination, veterinary care and prevention of diseases. Organic goat production has modestly increased or fluctuated in Italy and Spain, with France maintaining a fairly stable number since 2001. Percentage of goats in organic production is the highest in Italy, about 9%, followed by Greece, France and Spain (Eurostat, 2007).

Housing, handling, transport, slaughtering, disease, injury, starvation and veterinary treatment could be major considerations for organic goat producers to address consumers' concern for animal welfare. The standards for animal welfare in organic goat production are yet to be defined. Lindqvist (2001).

Sheep have been raised for their milk for thousands of years. Today the commercial dairy sheep industry is concentrated in the European and Midwestern countries on or near the Mediterranean Sea. France alone has almost one million ewes in dairy production. Most of the world's sheep milk is processed into cheese. Roquefort, the blue cheese of south central France, is one of the better known of the sheep milk cheeses (E. Molina Alcaide, 1995).

There is no obvious difference in production efficiency or carcass quality between organic and conventional goat herds. Guzmán et al. (2008) used 24 identical twins and studied the carcass yield in Blanca Andaluza kids from birth to 8.4 kg. They reported differences in leg compactness index, offal distribution, but no difference in carcass yield and conformation.

Zurita et al. (2008) studied 89 Murciano-Granadina kids from birth to 7 kg and reported a 0.5 kg difference between the conventional and organic herd. Organic goat meat production generally favors a feeding system that avoids a high energy diet with maximum weight gain. As a result, the carcass will likely accumulate less amount of internal fat around their heart and kidneys. A slower rate of gain, but heavier market weight, is therefore more advisable.

Some feeds can impart undesirable flavors to the milk (e.g. fish meal) and should not be fed in large quantities during lactation. Copper can be toxic to sheep. Only mineral supplements and prepared feeds that have been specially formulated for sheep should be fed, because many feeds for other livestock species contain added copper. Specially formulated mineral mixes for sheep also contain added selenium which is required by sheep but deficient in much of the Great Lakes' region (Dawson, L.J., Sahlu, T., 2008).

It has been demonstrated that organic dairy goat production can be productive and sustainable. Rahmann (2002) reported an annual average milk yield from 532 to 835 g/day of a herd with 20–35 lactating goats over a period of 6 years in Germany. However, the availability of forage, affected by rainfall and other environmental factors, contributed to the fluctuation in milk yield over a 6-year period. In a more recent study/survey with a larger herd over a 4-year period, Rahmann (2008) reported a milk yield ranging from 488 to 790 kg/year, or 1.6 to 2.6 kg/day for a 305-day lactation period. These reports established that higher yields are possible in organic dairy goat production, but is subject to pasture condition.

Moroni et al. (2002) studied intramammary infections, milk production and quality in organic dairy goat farming and concluded that it is possible to control infections and somatic cell counts without the use of drugs and maintain production level and quality.

5. Alternative Feed protein Resources and Their Effects on the Quality of Meat and Milk from Small Ruminants

Literature on the use of alternative legume seeds in animal nutrition is copious, covers a long time period and has caused renewed interest in recent years.

Some Mediterranean countries are characterized by harsh climate conditions. In these regions, pasture is available only for short periods or is not available at all. Moreover, the use of cereals in animal diets creates a competitive conflict with human nutrition, and the use of soybean is expensive. An interesting challenge for scientists in the field of animal nutrition is the introduction of alternative feedstuffs that could overcome the problems of environmental harshness and production costs. At the same time, the preservation of animal health, production yield and product quality is essential (European Commission, 2005).

The quality of meat and milk from sheep and goats offered alternative feeds as a replacement for concentrates. Legume seeds and pods, shrubs, local agro-industrial by-products or novel pasture species are cheap and widely available in Mediterranean countries and are suitable for sheep and goat nutrition. Many of these alternative feed resources (AFR) contain secondary compounds, such as tannins. Tannin-containing feeds result in meat of a lighter colour and tend to increase milk yield and protein content, probably because they protect dietary proteins from ruminal degradation (Makkar, 2003; Min et al., 2003).

Conjugated linoleic acid (CLA) content in kid meat can be increased by feeding animals chopped cactus cladodes. Grazing saltbush (Atriplex spp.) preserves lamb meat colour stability, suggesting that the high level of vitamin E in these shrubs protects myoglobin from oxidation. When olive cake silage is included in lamb or ewe diets, linoleic and oleic acid contents may increase in meat and milk fat, respectively. The appearance of terpenes in sheep and goat milk is enhanced by grazing on some novel pasture species, such as *Galium verum*, *Cichorium intybus* and *Chrisantemum coronarium*, which modify milk and cheese sensorial profile, compared to grazing on conventional forages. (H. Ben Salem, A. Priolo, 2001).

The ban of meat meals due to the Bovine spongiform encephalopathy (*BSE*) crisis and public concern about the widespread use of GMO-feedstuffs, such as soybean, have heightened attention towards the use of local home-grown vegetable protein sources, such as alternative legume grains, to satisfy animal protein requirements. Moreover, the introduction of EEC regulation for organic livestock farming (Council Regulation European Economic Community (EEC), 1804/1999) has supported the use of local legume crops due to their positive ecological role in reducing soil nitrogen depletion and breaking pest and disease cycles (Caballero, 1999).

The use of field beans (Vicia faba) in varying proportions, in replacement of soybean meal and sunflower meal, did not affect lamb performance (Caballero et al., 1992).

In a study on the total replacement of soybean cake with various legume seeds (lupins, lentils or faba beans), breed Aragonesa lambs fed faba beans showed higher weight gains than those fed lupins or lentils (Purroy et al., 1992a). Moreover, carcasses from lambs fed faba beans showed lower fatness than those from lambs fed soybean cake or lentil diets. Lanza et al.(1999) observed that the growth performances of lambs fed a faba bean-rich diet (540 g/kg,as fed) in total replacement of soybean meal, were similar to those of lambs fed soybean meal. Nevertheless, the faba bean-fed lambs had a lower lean deposition compared to the soybean-fed animals. Besides, meat from lambs fed faba bean diets showed better results on sensory analysis, leading to a higher acceptability compared to the meat of animals offered the soybean diet.

The partial or total replacement of soybean meal with peas (*Pisum sativum*) did not affect the growth performance of Aragonesa lambs (Purroy and Surra, 1990). Purroy et al. (1992a) demonstrated an increase in digestibility of dry matter as the percentage of *peas* in lamb diets increased.

Loe et al. (2004) studied the energy value and the optimum inclusion level of field peas in corn-based diets supplied to finishing lambs and concluded that field peas could suitably replace corn, up to high levels (450 g/kg DM), and had an energy density comparable to that of corn.

Lanza et al. (2003a) reported that the inclusion of 180 or 390 g/kg (as fed) of peas into lamb diet did not affect meat ultimate pH, colour, or cooking losses, while it caused higher meat drip losses than the soybean meal diet. The replacement of soybean meal and barley grain by chickpeas (*Cicer arietinum*) did not affect fattening performances in Chios lambs (Hadjipanayiotou, 2002) and in Florina (Pelagonia) lambs (Christodoulou et al., 2005). Lanza et al. (2003b) and Priolo et al. (2003) also investigated the effects of the inclusion of chickpeas in partial or total replacement of soybean meal in the diet on lamb meat quality. Meat ultimate pH, colour, drip and cooking losses, as well as shear force, were not affected by dietary treatments (Lanza et al., 2003b).

Intramuscular fatty acids (FA) composition of the meat of chickpea-fed lambs showed a higher (CLA) content than that of soybean-fed lambs. Moreover, the level of C22:5 n-3 and the sum of n-3 fatty acids were higher in the meat from lambs fed a diet with a high proportion of chickpeas (420 g/kg, on as fed basis) than in that from lambs fed lower amounts (200 g/kg, on as fed basis) of chickpeas or soybean meal (Priolo et al., 2003).

Narbon beans (*Vicia narbonensis*) supplemented at different levels (0.01 and 0.02 of body weight) to pasture hay-fed lambs increased animal weight gain compared to similar levels of supplementation of field peas (*Vicia sativa*) (Jaques et al., 1994).

Yu et al. (2001) compared the meat flavour of lambs fed raw or roasted narbon beans or lucerne. Sensory evaluation showed that the "flavour strength" and "aroma strength" of the meat from lambs fed roasted or raw narbon beans were higher than those of the meat from lambs fed lucerne. Nevertheless, the scores for the overall acceptability of meat from lambs fed the narbon bean diets was comparable to that of the lambs fed the lucerne diet.

Production of meat and milk is influenced by both the reproductive efficiency of the herd and the production levels achieved by the animal being fattened, while flavour and tenderness are evaluated during eating.

It is well known that animal dietary regimen strongly affectsmeat colour (Priolo et al., 2001). and the FA composition of meat (Enser et al., 1998;Wood et al., 2003) and milk (Chilliard et al., 2000; Chouinard et al., 2001) as well as their flavour (Urbach,1990; Bendall, 2001; Vasta and Priolo, 2006). Moreover, some volatile compounds, such as terpenes, derived from pastures and transferred to meat, milk and dairy products, can be used as tracers of sheep and goat feeding preferences (Priolo et al., 2004; Fedele, 2005) or geographical origin (Viallon et al., 2000).

EXPERIMENTAL PART

Introduction

The need to meet the demand of dairy products from the point of view of the guaranteed health security and high quality nutritional and organoleptic quality is enhanced by time bands growing number of consumers.

In fact, milk and cheese produced in extensive systems, respecting the environment and the animal itself, from subjects of local breeds raised primarily for grazing, possibly supplemented with grains in farm produced, gained considerable value and appreciation in the market.

In this context fits the current need to substitute in animal feed, soybeans and corn for food alternative, depending not only of their high market price, but also the risks associated with GMOs and the presence of possible contamination mycotoxin.

As for corn, an energy source that most commonly found in commercial foods, in Mediterranean environments and animals with high nutritional requirements does not, it is replaced with barley, corn can find it easily in Sicily.

As for the replacement of soybean, grain legumes, traditionally cultivated in Sicily, respond well to this need, both for their willingness at the local level, both being GM - free and therefore more nutritionally safe.

The choice of the grain legumes used as an alternative to soy protein and energy for the integration of fodder must surely be based on agronomic and production of the crop, but must also take account of the impact on production levels and quality requirements are the same able to give the products.

Recent studies on the use of grain legumes for sheep tag in larger measure the production of meat (Wide et al., 2008).

These studies have called attention to the effects of type of concentrate protein on carcass composition and organoleptic and sensory characteristics of sheep. Present a few experiences that are specific to the study of the effects of different pulses on the production of milk and dairy sheep.

In times of shortage of pasture or high nutritional requirements of lactating ewes, the grains may be an appropriate complement of legume forage based ration, replacing the feed trade, sometimes too much protein, this will reduce the waste of economic and nitrogen emissions into the environment, but also prevents the rise in blood and milk of animals to which it is bound, as well as the deterioration in the attitude of dairy milk, the onset of disease in the breast and limbs and reproductive problems (Guo et al., 2004; Mellado et al., 2004; Mitchell et al., 2005).

Moreover the use of grain legumes in animal feeding obtained locally, which is also applicable in systems operating under organic production, could contribute to the spread and development of the cultivation of their crop, acting as a driving force for the increase off farm income of cereal-forage-livestock enterprises, today more and more difficulty breathing, and also making a contribution to the enhancement of sheep dairy products deriving from it, relying on the nutritional and health of themselves.

The objective of this thesis is therefore to provide a contribution to the knowledge of the chemical-nutritional few grains of legumes grown locally to be used as protein sources for feeding of sheep as an alternative to soy milk, and get directions on the legume species in order to pursue, along with good production levels, better diet and organoleptic characteristics, as well as a high security level nutrition of dairy sheep. To this end, we evaluated the effects of the food administration of chickpea, Faba bean or pea seeds, mixed with barley, production and quality of milk obtained from lactating ewes.

Materials and methods

The research was conducted in 2007 at the company Pietranera Foundation A. and S. Lima-Mancuso, University of Palermo, situated in the territory of S. Stefano Quisquina (AG) (37° 30'N, 13° 31'E, 178 m), areal representative to climatic conditions, and typical Sicilian hinterland hills of the breed of sheep farming Comisana and production of Pecorino Siciliano cheese.

The food analysis, carried out before the test to ensure a correct formulation of rations, involved the determination of dry matter, crude protein (CP), ether extract (EE) and ash (AOAC, 1990), NDF (Goering and Van Soest, 1970), ADF and ADL (Van Soest and Robertson, 1980). The percentage content in non-structural carbohydrates (CNS) was calculated by subtracting the value at 100 percent of the content in crude protein, ether extract, ash and NDF. The energy value of food was expressed as net energy for lactation (EN_L), calculated on the basis of digestibility and estimated using the equation proposed by Van Soest and Fox (1992).

The chemical composition of hay and concentrates that make up the experimental diets is shown in Table 1.

Based on the analysis were then formulated concentrates with the same nitrogen (3.7% nitrogen on DM), to be administered to the sheep being tested in a similar amount (800 g / head / day divided into two meals daily) according to the following schedule:

a) 500 g + 300 g chickpeas and barley;

b) 450 g + 350 g broad bean and barley;

c) 550 g + 250 g pea and barley;

d) 800 g of commercial concentrate.

The main components of commercial concentrate, which represents the control, are made from conventional sources, such as corn and soybean meal.

The forage base of the diet of all animals consisted of hay-based company clover and wild grasses used in the will, and then the diet is only differentiated by source of protein concentrate.

For this experiment, were used 8 Comisana breed ewes, divided into pairs homogenized according to the days of lactation (92 ± 9 d) of live weight (55.9 ± 6.3 kg) of body condition (BCS = body condition score of 2.87 ± 0.18) and milk production (875 ± 60 g / d).

After a suitable period of adaptation to housing conditions and the food rations, the test was started which lasted eight weeks. The experiment was divided into four phases of two weeks in each of which each of the four pairs of sheep to take advantage of a different focus, according to the format of a 4x4 Latin square design.

Each sheep used in experiments take advantage of single garage and manger to allow the individual administration of hay and concentrate.

Daily was measured for each ewe, milk production, the amount of concentrate and hay administered and residuals.

At the beginning and end of each phase, or a biweekly basis, we proceeded to the measurement of live weight and BCS of ewes.

3 controls were performed for each of the 4 experimental stages Week, respectively on days 10, 12 and 14 where the samples were run: of hay and concentrates (given the individual and any residue), and individual milk.

The individual milk samples were analyzed for fat, protein, casein, lactose, somatic cell count (Milkoscan FT 6000, Foss Electric, Hillerød, Denmark) and urea (CL-10 Plus, Eurochem, Italy), technological parameters as r (clotting time, min), k20 (speed of clot formation, min) and a30 (Curd firmness, mm) were determined with Formagraph instrument (Foss, Italy).

The normalized milk (LN) with 6.5% fat and 5.8% of protein was calculated using the following formula proposed by Pulina and Nudda (2002): LN = kg milk * (0.25% + fat + protein %).

The data were statistically analyzed using the MIXED procedure of SAS 9.1.2 software (2004) using a model that considered the dietary treatment as fixed effect and the experimental phase and sheep as random effects.

The values of somatic cells (CCS) and total bacterial count (CMT) have been reported in logarithmic form (log 10). Differences between means were tested with the "t" test of Student and reported below the threshold of 0.05.

Results and Discussion

In table 2 shows the food intakes of animals recorded during the test, divided by the concentrated and used for hay. Not all concentrates, except the one with the field bean, have always been totally consumed by sheep, if we admit that this is dependent to a greater extent than their palatability, the test concentration, and the one containing the field bean, in particular, are definitely more pleasing results than the commercial feed to sheep. It had recorded the lowest consumption and, therefore, a ratio of concentrate in the diet lower by about 5% compared to the test concentrations.

There were no significant differences, however, in terms of total intake (hay more concentrated) between the different experimental groups of the test, which means that the sheep who received commercial feed offset by eating a greater quantity of hay. For these reasons, then, considering the total ingestion by animals, significant differences were found between the groups in terms of ingestion of non-structural and structural carbohydrates, due to the fact that the group fed the commercial feed has ingested a larger amount of fiber (NDF, ADF, ADL) and fewer CNS.

Differences, although these have not reached the threshold of significance, have emerged in terms of ENL ingested in the diet, as with commercial feed were recorded lower values than other concentrates. Consequently, the ratio PG / ENL animals fed the commercial feed showed higher values than the other groups.

Animals in the group receiving the chickpea have ingested both the concentrate with hay that a quantity of fiber than the other, with consequent impact on increasing the ratio PG / NDF of the diet ingested as shown in table 2.

The sheep receiving concentrates with pea and faba bean produced significantly greater milk than the animals fed with chickpea, intermediate levels were found by the group fed with commercial feed. While Table 3 shows data on the amount and quality of milk produced by animals during the test.

1 = milk normalized to 6.5% fat and 5.8% protein percentage
2 = grams of milk protein / kg of dietary protein ingested
3 = grams of milk casein / kg of dietary protein ingested

The production differences between the various groups are rather insignificant in terms of standardized milk. The use of standardized milk parameter, which considers the quality of milk with regard to the percentages of protein and fat, which provide the cheese yield, it is certainly important in sheep than other species because, as is known, the sheep's milk cheese is launched at all and is not intended for fresh consumption.

The flattening of the productive differences between the groups occurred for the standardized milk is due to the higher fat content, however, found in milk of the group fed chickpea than the others.

At the level of nitrogenous components of milk, were found in protein and casein levels slightly higher, although not significantly, with the test concentration, while the level of urea was within normal limits with all diets.

There were no differences between groups for total bacterial count and the somatic cell count milk.

The parameters of milk coagulation (r, K20 and a30) were similar between the groups with values and satisfactory, except for the clotting time results for all animals, slightly higher than normal.

Interesting results emerge from the analysis of data on the efficiency of protein utilization for the synthesis of protein and milk casein. Significantly more favorable values were obtained for animals that benefited from the concentrates containing pea and faba bean.

This shows that the nitrogen content of these legumes are used more favorably by the ewes, probably because of their amino acid composition or even better than their degradability in the rumen that favor synchrony between energy and protein available for the synthesis of microbial protein.

In terms of live weight and body condition score, all sheep showed some improvement between the beginning and end of each phase of food. However, this improvement was more content in sheep that received the chickpea, and these, in fact, despite their lower milk production, showed a lower weight gain of only 0.5%, and only a subtle increase in the body condition score. All weights and BCS analysis showed no significant differences between groups, with the exception of the BCS at the end of phase food, higher for sheep fed faba beans compared with those receiving the chickpea.

Conclusion

Ultimately, based on results, the use of concentrated preparations in company with grains of cereals and legumes available locally does not lead to deterioration in quantity and quality characteristics of milk production compared to balanced compound feed trade which have been made corn and soybeans, even in some cases these are improved.

In comparing the three protein sources did not reveal significant differences, although the broad bean and pea seeds seem to allow a greater milk production and higher production efficiency of protein and milk casein with the same dietary protein ingested.

Tables

Table 1. Chemical composition of foods used in the test (% s.s.)

	Barley	Chickpea	Faba been	Pea	Feed	Hay
Dry Matter	89,02	90,45	86,75	86,92	88,04	91,68
Crude protein	13,38	25,65	27,59	24,57	23,31	12,14
NDF	17,21	12,40	19,33	19,98	23,15	54,79
ADF	9,03	5,50	12,60	8,95	10,97	39,91
ADL	1,03	0,28	0,58	0,18	2,60	5,29
Ether extract	1,76	5,09	1,39	1,48	5,09	1,67
Ash	2,98	3,56	3,35	3,16	9,22	10,74
CNS	64,67	53,30	48,34	50,81	39,23	20,66
EN_L Mcal/kg s.s.	2,02	2,20	2,03	2,10	1,80	1,27

Table 2. Feed intake during test

			Chickpea	Faba been	Pea	Feed	P <
Concentrate	Concentrate	g/d t.q.	780 a	800 a	774 a	667 b	***
	Barley	% t.q.	36,5 b	43,7 a	29,5 c		***
	Concentrate	g/d s.s	702 a	702 a	678 a	587 b	***
	Barley	% s.s.	36,2 b	44,4 a	30,0 c		***
	Crude protein	g/d	150 a	151 a	144 ab	137 b	+
	NDF	g/d	102 b	133 a	132 a	136 a	***
	ADF	g/d	47,7 c	77,3 a	60,9 b	64,4 b	***
	ADL	g/d	3,91 c	5,47 b	2,99 c	15,3 a	***
	Ether extract	g/d	27,1 b	10,9 c	10,6 c	29,9 a	***
	Ash	g/d	23,5 b	22,4 b	21,1 b	54,1 a	***
	CNS	g/d	401 a	387 ab	371 b	230 c	***
	EN_L	Mcal/d	1,50 a	1,42 a	1,41 a	1,06 b	***
Hay	Dry Matter	g/d	1619 c	1741 ab	1681 bc	1809 a	***
	Crude protein	g/d	220 b	230 b	228 b	240 a	***
	NDF	g/d	842 c	912 ab	874 bc	946 a	***
	ADF	g/d	618 c	670 ab	642 bc	693 a	***
	ADL	g/d	89,2 b	94,7 a	93,5 ab	97,3 a	*
	Ether extract	g/d	30,6 b	31,7 b	31,7 b	33,4 a	***
	Ash	g/d	171 c	185 ab	179 bc	191 a	***
	CNS	g/d	356 c	384 ab	368 bc	399 a	***
	EN_L	Mcal/d	2,05 c	2,21 ab	2,12 bc	2,32 a	***
Diet	Hay	% s.s.	69,6 b	70,7 b	70,7 b	75,2 a	***
	Concentrate	% s.s.	30,4 a	29,3 a	29,3 a	24,8 b	***
	Dry Matter	g/d	2320	2443	2359	2396	ns
	Crude protein	g/d	370	380	372	377	ns
	NDF	g/d	944 c	1044 ab	1006 b	1082 a	***
	ADF	g/d	666 b	748 a	703 b	757 a	***
	ADL	g/d	93,1 c	100 b	96,5 bc	113 a	***
	Ether extract	g/d	57,7 b	42,7 c	42,4 c	63,3 a	***
	Ash	g/d	194 b	207 b	200 b	245 a	***
	CNS	g/d	756 ab	771 a	739 b	630 c	***
	EN_L	Mcal/d	3,55 a	3,64 a	3,53 a	3,37 b	ns
	PG/NDF		0,41 a	0,37 b	0,38 b	0,36 b	***
	PG/EN_L		106 b	105 b	106 b	112 a	***

a,b,c = P≤0,05; *** = P≤0,0001* = P≤0,05; + = P≤0,10; ns = not significant

Table 3. Production, composition and rheological parameters of the individual milk

		Chickpea	Faba been	Pea	Feed	P <
Milk	g/d	654 b	710 a	718 a	677 ab	**
Standardized Milk	g/d	735	766	781	741	ns
Fat	%	7.66 a	7.11 b	7.22 b	7.29 ab	*
Protein	%	6.74	6.86	6.73	6.58	ns
Casein	%	5.24	5.28	5.17	5.09	ns
Casein N / total N		77.6 a	77.0 bc	76.8 c	77.2 b	***
Urea	mg/dl	37.6	38.8	37.5	38.3	ns
Lactose	%	4.32	4.29	4.29	4.35	ns
CCS	$\log_{10}$ /ml	6.00	5.81	5.76	5.98	ns
CMT	$\log_{10}$ /ml	5.22	5.37	5.15	5.61	ns
Treatable acidity	SH/50	4.94	4.72	5.08	4.72	ns
pH		6.61	6.59	6.59	6.63	ns
R	min	26.8	26.5	28.2	28.5	ns
k20	min	1.80	1.82	2.68	1.91	ns
a30	mm	47.2	53.1	48.4	41.4	ns
Efficiency of protein synthesis [2]		115 b	125 a	130 a	114 b	*
Efficiency of casein synthesis [3]		89.1 b	96.3 a	99.4 a	88.4 b	**

a,b,c = P≤0,05; *** = P≤0,001;** = P≤0,01;* = P≤0,05; ns = not significant

Table 4. Weight and body condition (BCS) of animals at the beginning and end of trial period

		Chickpea	Faba been	Pea	Feed	P <
Initial weight	kg	58.8	58.4	57.7	58.9	ns
Final weight	kg	59.1	60.2	59.8	60.0	ns
Increased weight	%	0.53	3.04	3.59	2.00	ns
Initial BCS		2.94	2.97	2.91	3.00	ns
Final BCS		2.97 b	3.12 a	3.03 ab	3.06 ab	+
Increased BCS		0.03	0.15	0.12	0.06	ns

a,b = P≤0,05; + = P≤0,10; ns = not significant

Bibliography

AOAC, 1990. Official methods of analysis. 15th edn. Association of Official Analytical Chemist, Washington DC.

Anderson, V. L. 1999. Field peas in diets for creep feed for beef calves. Pages 1-4 in Carrington Research Extension Center Beef Production Field Day Report. Vol. 22. North Dakota State Univ., Fargo.

Boyazoglu, J., Morand-Fehr, P., Mediterranean dairy sheep and goat products and their quality. A critical review. Small Rumin. Res. 40, 1–11, 2001.

Bach, A., S. Calsamiglia and M. D. Stern. 2005. Nitrogen metabolism in the rumen. J. Dairy Sci. 88 (E. Suppl.) E9-E21.

Becker, D, E., L. J. Hanson, A. H. Jensen, S. W. Terrill and H. W. Norton. 1956. Dehydrated alfalfa meal as a dietary ingredient for swine. J. Anim. Sci. 15:820-829.

CC-IEE 2008 Report of the independent external evaluation of the Food and Agriculture Organisation of the United Nations (FAO). Rome, Italy: Food and Agriculture Organisation.

Caballero, R., Rioperez, J., Fernandez, E., Marin, M.T., Fernandez, C., 1992. A note on the use of field beans (Vicia faba) in lamb finishing diets. Anim. Prod. 54, 441–444.

Cubero, J.I., 1983. Origin,evolution and genetic improvement of grain legumes. In:Cubero, J.I., Moreno, M.T. (Eds.), Legumes for Grain, Mundiprensa, Madrid, Spain.

Caballero, R., 1999. Castile-La Mancha: a once traditional and integrated cereal-sheep farming system under change. Amer. J. Alt. Agric. 14, 188–192.

Calsamiglia, S., A Ferret, and M. Devant. 2002. Effects of pH and pH fluctuations on microbial fermentation and nutrient flow from a dual-flow continuous culture system. J. Dairy Sci. 85:574-579.

Delgado, C., Rosegrant, M., Steinfeld, H., Ehui, S. & Courbois, C. 1999 Livestock to 2020—the next foodrevolution. Food, Agriculture and Environment Discussion Paper 28. Rome, Italy: IFPRI, FAO and ILRI.

Delgado, C. 2005 rising demand for meat and milk in developing countries: implications for grasslands-based livestock production. In Grassland: a global resource (ed.D. A. McGilloway), pp. 29–39. The Netherlands: Wageningen Academic Publishers.

Duffus, C.M., Duffus, J.H., 1991. Introduction and overview. In: D' Mello, J.P., Duffus, CM., Duffus, J.H. (Eds.), Toxic Substances in Crop Plants. The Royal Society of Chemistry, Cambridge, UK, pp. L-20.

Enver Doruk Özdemir, Marlies Hardtlein, Ludger Eltrop, Land substitution effects of biofuel side products and implications on the land area requirement for EU 2020 biofuel targets, Energy Policy, Volume 37, Issue 8, August 2009, Pages 2986-2996, ISSN 0301-4215, DOI: 10.1016/j.enpol.2009.03.051.

Eurostat, 2007. Statistical Office of the European Communities (http://epp.eurostat.ec.europa.eu).

European Commission, 2005. Organic Farming in the European Union: Organic Farming in the European Union Facts and Figures. Director General of Agriculture and Rural Development. Brussels, November 3, p. 30.

Edwards R., J.F. Larivé, V. Mahieu and P. Rouveirolles, Well-to-wheels analysis of Future automotive fuels and powertrains in the European context 2006 update.
EUCAR/CONCAWE/JRC, Brussels (2006).
FAO 2009a Food balance sheets: FAOSTAT. FAO 2009b How to feed the world in 2050. Rome, Italy: Food and Agriculture Organization.

Gallagher, The Gallagher review of the indirect effects of biofuels production, Renewable Fuels Agency (July 2008).

Garcı 'a, M. A., J. F. Aguilera, and E. Molina Alcaide. 1995. Voluntary intake and kinetics of degradation and passage of unsupplemented and supplemented pastures fromsemiarid lands in grazing goats and sheep. Livest. Prod. Sci. 44:245–255.

Graham P, Vance CP (2003) Legumes: importance and constraints to greater use. Plant Physiol 131: 872–877.

Gatel, F., 1994. Protein quality of legume seeds for non-ruminant animals: A literature review. Anim. Feed Sci. Tech. 45:317–348.

Gallo, A., Moschini, M., Masoero, M., 2008. Aflatoxins absorption in the gastro-intestinal tract and in the vaginal mucosa in lactating dairy cows. Ital. J. Anim. Sci. 7:53-63. Ital. J. Anim. Sci. 7:53-63. Giorni, P., Battilani, P., Pietri, A., Magan, N.

Guzmán, J.L.,Delgado-Pertí ̃nez,M., Zarazaga, L.A., Flores,A., Puerta, R., Celi, I., Acosta, J.M., Argüello, A., 2008. Effect ofmanagement systemon the regional composition and offal distribution in Blanca Andaluza goat kids. In: Proceedings of the 9th International Conference on Goats, Queretaro, Mexico, August 31–September 4, pp.Goering H.K., Van Soest P.J., 1970. Forage fiber analysis. Agricolture handbook, 379.

Guo, K., Russek-Cohen, E., Varner, M.A., Kohn, R.A., 2004. Effects of milk urea nitrogen and other factors on probability of conception of dairy cows. J. Dairy Sci. 87:1878-1885.

Hoerr, F. J. 1998. Pathogenesis of enteric diseases. Poult. Sci. 77:1150–1155.

Kommera, S. K., R. D. Mateo, F. J. Neher, and S. W. Kim. 2006. Phytobiotics and organic acids as potential alternatives to the use of antibiotics in nursery pig diets. *Asian-australas J. Anim. Sci.* **19**:1784.

Khasan, A.M., T.K. Tashev, N.A. Todorov, and A.M. Hasan. 1989. Lucerne haylage, sunflower meal and peas as protein feeds in diets for dairy cows. Zhivotnov dniNauki 26(3):30-36.

Linn, J. G., N. P. Martin, W. T. Howard, and D. A. Rohweder. 1987. Relative feed value as a measure of forage quality. MN Forage Update. Vol. XII, No. 4 (Late Summer).

Lu, C.D.,Gangyi,X., 2008.Organic sheep and goat production. In: Presented at Annual Meeting of Chinese Sheep and Goat Association, Zinben, Shannxi, China, July 22–25 (Invited Paper).

Lecce, J. G. 1986. Diarrhea: The nemesis of the artificially reared, early weaned piglet and a strategy for defence. J. Anim. Sci.63:1307–1313.

Landblom, D.G, and W.W. Poland. 2000. Application of salt-limited pea/wheat midds creep diets in Southwestern North Dakota. Pp 8-12 in North Dakota Beef Cattle Range and Research.

Lu, C.D.,Gangyi,X., 2008.Organic sheep and goat production. In: Presented at Annual Meeting of Chinese Sheep and Goat Association, Zinben, Shannxi, China, July 22–25 (Invited Paper).

Lanza, M., Priolo, A., Biondi, L., Bella, M., Ben Salem, H., 2001. Replacement of cereal grains by orange pulp and carob pulp in faba bean-based diets fed to lambs: effects on growth performance and meat quality. Anim. Res. 50, 21–30.

Lanza,M., Pennisi, P., Priolo, A., 1999. Faba bean as an alternative protein source in lamb diets: effects on growth and meat quality. Zoot. Nutr. Anim. 25, 71–79.

Lindqvist, A., 2001. Animal health and welfare in organic sheep and goat farming experiences and reflections from a Swedish outlook. ActaVet. Scand. Suppl. 95, 27–31.

Loe, E.R.,Bauer,M.L., Lardy,G.P.,Caton, J.S.,Berg, P.T., 2004. Field pea (Pisumsativum) inclusion in corn-based lamb finishing diets. Small Rum. Res. 53, 39–45.

Mena, Y.,Nahed, J., Ruiz, F.A., Castel, J.M., Ligero,M., 2008.Goat production systems in mountainous areas: approach to the organic model. In: Proceedings of the 9th International Conference on Goats, Queretaro, Mexico, August 31–September 4, p. 97.

Mao, X. F., X. S. Piao, C. H. Lai, D. F. Li, J. J. Xing, and B. L. Shi. 2005. Effects of glucan obtained from the Chinese herb Astragalus *membranaceus* and lipopolysaccharide challenge on performance, immunological, adrenal, and somatotropic responses of weanling pigs. *J. Anim. Sci.* **83**:2775–2782.

Moroni, P., Bronzo, V., Cuccuru, C., Luzi, F., Cattaneo, D., Savoini, G.,2002. Organic dairy goat farming: intramammary infections, milk production and quality. Organic meat and milk from ruminants. In: Proceedings of a Joint International Conference Organised by the Hellenic Society of Animal Production and the British Society of Animal Science, Athens, Greece, 4–6 October 2001.

Makkar HPS (2003). Effects and fate of tannins in ruminant animals, adaptation to tannins, and strategies to overcome detrimental effects of feeding tannin-rich-feeds. Small Rumin. Res. 49: 241-256.

Mellado, M., Valdez, R., Lara, L.M., García, J.E., 2004. Risk factors involved in conception, abortion, and kidding rates of goats under extensive conditions. Small Ruminant Res. 55:191-198.

Mitchell, R.G., Rogers, G.W., Dechow, C.D., Vallimont, J.E., Cooper, J.B., Sander-Nielsen, U., Clay, J.S., 2005. Milk urea nitrogen concentration: heritability and genetic correlations with reproductive performance and disease. J. Dairy Sci. 88:4434-4440.

Peeters, E., B. Driessen, and R. Geers. 2006. Influence of supplemental magnesium, tryptophan, vitamin C, vitamin E, and herbs on stress responses and pork quality. *J. Anim. Sci.* **84**:1827–1838.

Puchala,R.,Animut,G.,Goetsch,A.L.,   Patra,A.K.,   Sahlu, T.,Varel,V.H.,Wells,J., 2008. Ruminal methane emissions by goats consuming dry hay of condensed tannin containing. In: Proceedings of the 9th International Conference on Goats, Queretaro, Mexico, August 31–September 4, p. 98.

Purroy, A., Surra, J., Mu˜ noz, F., Morago, E., 1992a. Empleo de leguminosas grano en el el pienso para cebo de corderos: III. Guisantes (Use of seed crops in the fattening diets for lambs: III. Pea seeds). Producci´ on Animal88A, 63–69 (in Spanish).

Purroy, A., Surra, J., Mu˜ noz, F., Morago, E., 1992a. Empleo de leguminosas grano en el el pienso para cebo de corderos: III. Guisantes (Use of seed crops in the fattening diets for lambs: III. Pea seeds). Producci´ on Animal 88A, 63–69 (in Spanish).

Purroy, A., Surra, J., 1990. Empleo de guisantes y de habas en el pienso para cebo de corderos (Use of peas and beans in feeds for fattening lambs). Archivos de Zootecnia 39, 59–66 (in Spanish).

Pulina G,. Nudda A., 2002. Milk production. In "Dairy sheep feeding and nutrition" (ed. G. Pulina), 11-27. Avenue media, Bologna, Italy.

Ribeiro, JL; NUSSIO, LG; MOURÃO, GB Effects of moisture absorbents, and chemical and microbial additives on OVALOR nutritional, and fermentation losses in silage marandugrass. Journal of Animal Science, v.38, n.2, p.230-239, 2009.

Ravindran V, Blair R (1992). Feed resources for poultry production in Asia and the Pacific II. Plant protein sources. World Poult. Sci., 48: 205-231.

Rahmann, G., 2002. On farm organic dairy sheep and goat breeding in Germany. In: Proceedings of the 14th IFOAMOrganicWorld Congress, p. 94.

Rahmann, G., 2008. Goat milk production under organic farming standards. In: Proceedings of the 9th International Conference on Goats, Queretaro, Mexico, August 31–September 4, p. 109.

SAS, 2004. SAS/STAT Qualification Tools User's Guide (version 9.1.2). Statistical Analysis System Institute Inc., Cary, NC, USA.

Todorov, N.A., 1988. Cereals, pulses and oilseeds. In: De Boer, F., Bickel, H. (Eds.), Livestock Production Science. Elsevier, Amsterdam, pp. 47-95.

Van Soest PJ, Fox DG (1992). Discounts for net energy and protein – fifth revision. Proceed. of Cornell Nutrition Conference, 40-68, Ithaca, New York.

Van Soest P.J., Robertson, J.B., 1980. Systems of analysis for evaluating fibrous feed. In: Pigden W.J., Balch C.C., Graham M., (Ed.), Standardisation of Analysis Methodology for Feeds. IDRC Ottawa, 49-60.

Vasta V., Nudda A., Cannas A., Lanza M., Priolo A., 2008. Alternative feed resources and their effects on the quality of meat and milk from small ruminants. Animal Feed Science and Technology, 147, 223–246.

Wang, Z., Hart, S.P., Goetsch, A.L.,Merkel, R.C., Dawson, L.J., Sahlu, T., 2008.

Effects of protein supplementation on Haemonchus contortus infection in goats. In: Proceedings of the 9th International Conference onGoats, Queretaro, Mexico, August 31–September 4, p. 268.

Yuan, S. L., X. S. Piao, D. F. Li, S. W. Kim, H. S. Lee, and P. F. Guo. 2006. Effects of dietary Astragalus polysaccharide on growth performance and immune function in weaned pigs. *Anim. Sci.* **82**:501–507.

Zurita, P., Camacho, M.E., Pleguezuelos, J., Delgado, J.V., 2008. Organic vs.conventional herd effects on theweights and daily gains in Murciano-Granadina kids. In: Proceedings of the 9th International Conference on Goats, Queretaro, Mexico, August 31–September 4, p. 108.

Integrazione con granelle di leguminose per la produzione di latte ovino biologico

A. BONANNO, A. DI GRIGOLI, G. TORNAMBÉ, V. BELLINA, F. MAZZA, M.L. ALICATA

Dipartimento S.En.Fi.Mi.Zo., sezione di Produzioni Animali, Università di Palermo

Parole chiave: cece, favino, pisello, latte ovino biologico.

INTRODUZIONE - Nell'alimentazione degli animali allevati in biologico, l'integrazione con concentrati costituiti da semplici miscele a base di granelle di leguminose, coltivate in azienda o reperibili localmente, possono costituire una valida alternativa ai mangimi commerciali che, contenendo mais e soia biologici OGM-free, risultano costosi e non privi del rischio di contaminazione da micotossine. Gli studi sull'uso di granelle di leguminose nella dieta degli ovini riguardano in larga misura la produzione di carne[1,6] mentre, tra i pochi lavori sulla produzione di latte ovino biologico[2,4], manca un confronto diretto tra le specie leguminose più diffuse in ambiente mediterraneo. L'obiettivo della prova è quello di valutare gli effetti dell'integrazione con cece, favino o pisello proteico in miscela con orzo, come possibili alternative al mangime commerciale, sulla produzione e sulla qualità del latte ovino ottenuto in regime biologico.

MATERIALI E METODI - Sono state utilizzate 12 pecore di razza Comisana alloggiate in box individuali e suddivise in quattro gruppi omogenei per giorni di lattazione (92±9 d), peso vivo (56±6 kg) e produzione di latte (875±60 g/d). Ciascun gruppo è stato alimentato con fieno a volontà e, seguendo lo schema di un quadrato latino 4x4 per fasi di 14 d, con 800 g/d di uno dei seguenti concentrati isoazotati (3,7% di azoto sulla SS): a) 500 g di cece e 300 g di orzo; b) 450 g di favino e 350 g di orzo; c) 550 g di pisello proteico e 250 g di orzo; d) mangime del commercio contenente mais e soia. Negli ultimi 5 d di ciascuna fase, per ciascuna pecora sono state misurate e campionate la produzione di latte, le quantità somministrate e quelle residue di fieno e concentrato. Alla fine ed all'inizio di ogni fase si sono rilevati il peso vivo e il BCS delle pecore. Sugli alimenti sono stati determinati sostanza secca, proteina grezza, estratto etereo, ceneri e carboidrati strutturali. Il latte è stato analizzato per la determinazione di grasso, proteina, caseina, lattosio e cellule somatiche (CCS) (Combifoss 6000, Foss Italia), urea (CL-10 Plus, Eurochem, Italia) e dei parametri di coagulazione (Formagraph, Foss Italia), mentre è in corso di definizione la composizione in acidi grassi. L'elaborazione statistica dei dati è stata effettuata con la procedura MIXED del SAS 9.1.2[5], usando un modello in cui l'effetto fisso è il tipo di concentrato e gli effetti casuali sono la fase e la pecora. Le differenze fra le medie sono state testate con il test "*t*" di Student (P≤0,05).

RISULTATI E CONSIDERAZIONI - I tre concentrati contenenti le granelle di leguminose (Tabella 1) sono stati maggiormente consumati rispetto al mangime commerciale, risultando quindi di buona appetibilità per le pecore. Poiché l'ingestione totale di fieno e concentrato non si è differenziata tra i gruppi, le pecore che ricevevano il mangime commerciale hanno compensato consumando più fieno, da cui la più alta ingestione di NDF. La miscela con cece e il mangime, più dotati in lipidi (3,9, 1,6, 1,6 e 5,1% sulla SS, per cece, favino, pisello e mangime), hanno comportato una maggiore ingestione in estratto etereo. La produzione individuale media di latte è stata superiore con il pisello e il favino rispetto al cece, mentre livelli intermedi sono stati registrati con il mangime. Il grasso del latte ha mostrato un trend rapportabile, per l'effetto diluizione, con la quantità di latte prodotta piuttosto che con l'ingestione lipidica. I tenori in proteina, caseina e urea del latte non hanno mostrato variazioni per effetto del concentrato, mentre il rapporto tra azoto (N) caseinico e N totale è migliorato con il cece. Non sono state rilevate differenze fra i gruppi anche per il contenuto in cellule somatiche e per i parametri di coagulazione del latte (r, k$_{20}$ e a$_{30}$). L'efficienza di utilizzazione della proteina alimentare per la sintesi di caseina del latte ha mostrato valori più elevati per le pecore alimentate con pisello e favino. La migliore utilizzazione nutrizionale della componente azotata di tali leguminose potrebbe dipendere dalla loro migliore composizione aminoacidica, o anche dalla loro maggiore degradabilità[3] che favorirebbe nel rumine la sincronia tra energia e proteina disponibili per la sintesi di proteina microbica. Tutte le pecore hanno mostrato un incremento di peso vivo e un miglioramento di condizione corporea tra l'inizio e la fine di ciascuna fase alimentare, senza che vi fossero differenze fra i gruppi. In definitiva, i risultati ottenuti dimostrano come l'uso di concentrati preparati in azienda miscelando l'orzo

con granelle di leguminose reperibili localmente non comporti un peggioramento delle caratteristiche quanti-qualitative della produzione di latte rispetto ad un mangime del commercio contenente mais e soia. Nel confronto tra le tre fonti proteiche non sono emerse differenze notevoli, tuttavia favino e pisello proteico sembrano consentire, a parità di proteina alimentare ingerita, una maggiore produzione ed una più elevata efficienza della sintesi di caseina del latte.

Ricerca effettuata nell'ambito del progetto In.Fo.Pro. finanziato dall'Assessorato Agricoltura e Foreste della Regione Siciliana.

Tabella 1 - Effetto del tipo di concentrato sull'ingestione e sulla produzione di latte delle pecore.

	Cece	Favino	Pisello	Mangime
Ingestione, g/d				
SS del concentrato	702 a	702 a	678 a	587 b
SS totale	2320	2443	2359	2396
Proteina grezza	370	380	372	377
NDF	944 c	1044 ab	1006 b	1082 a
Estratto etereo	57,7 b	42,7 c	42,4 c	63,3 a
Produzione di latte				
Latte, g/d	654 b	710 a	718 a	677 ab
Lattosio, %	4,32	4,29	4,29	4,35
Grasso, %	7,66 a	7,11 b	7,22 b	7,29 ab
Proteina, %	6,74	6,86	6,73	6,58
Caseina, %	5,24	5,28	5,17	5,09
N caseinico/N totale	77,6 a	77,0 bc	76,8 c	77,2 b
Urea, mg/dl	37,6	38,8	37,5	38,3
CCS, log$_{10}$ 1000/ml	6,00	5,81	5,76	5,98
r, min	26,8	26,5	28,2	28,5
k$_{20}$, min	1,80	1,82	2,68	1,91
a$_{30}$, mm	47,2	53,1	48,4	41,4
[1]ESC	89,1 b	96,3 a	99,4 a	88,4 b

a,b,c = P≤0,05.
[1]ESC= efficienza sintesi caseina, in g caseina/kg PG ingerita.

■ Legume grains as dietary supplements in organic sheep milk production

Key words: chickpea, faba bean, pea, organic sheep milk.

Bibliografia

1. Bonanno A., Di Grigoli A., Tornambè G., Mazza F., Di Miceli G., Frenda A.S., Giambalvo D. (2009), Atti del IV Workshop GRAB-IT (Gruppo di Ricerca in Agricoltura Biologica - Italia) Ed. Università di Palermo, 311-316.
2. Di Grigoli A., Pollicardo A., Mele M., Bonanno A., Tornambè G., Vargetto D. (2009), Options Mèditerranèennes, Séries A: Mediterranean Seminars, 85, 459-464.
3. Hadjapanaioyu M. (2002). Animal Feed Sciece and Technology, 96, 103-109.
4. Molle G., Decandia M., Cabiddu A., Acciaro M., Fois N., Sitzia M. (2009), Proceedings of the 13°Seminar of FAO-CIHEAM Sub-Network on sheep and goat nutrition, Leon (Spain) 14-16 October 2009 (Abst).
5. SAS, 2004. SAS/STAT Qualification Tools User's Guide (version 9.1.2). Statistical Analysis System Institute Inc., Cary, NC, USA.
6. Vasta V., Nudda A., Cannas A., Lanza M., Priolo A. (2008), Animal Feed Science and Technology, 147, 223-246.